YOUR KNOWLEDGE HAS VALUE

- We will publish your bachelor's and
  master's thesis, essays and papers

- Your own eBook and book -
  sold worldwide in all relevant shops

- Earn money with each sale

Upload your text at www.GRIN.com
and publish for free

Bibliographic information published by the German National Library:

The German National Library lists this publication in the National Bibliography; detailed bibliographic data are available on the Internet at http://dnb.dnb.de .

Imprint:

Copyright © 2019 GRIN Verlag
Print and binding: Books on Demand GmbH, Norderstedt Germany
ISBN: 9783668947849

This book at GRIN:

https://www.grin.com/document/469110

Disha Kumari

In Vitro Anticancer studies on Colon rectum cell line

GRIN Verlag

"In Vitro Anticancer studies on Colon rectum cell line"

CONTENT

I. INTRODUCTION

Quercetin

Quercetin is scientifically known as 3,3',4',5,7-pentahydroxyflvanone which was given by IUPAC. Quercetin is derived from a latin word known as "Quercetum" that means oak forest. This comes under the class known as Flavanols that is not produced by human body. The quercetin is yellow in color that is poorly soluble in hot water, partially soluble in alcohol and is insoluble in cold water. It is most widely used Flavonoid worldwide to treat many disorders. It shows many biological activities like antiallergenic, antiviral, vasodilating and anti-inflammatory disorders, anti-platelet, anti-tumor, anti-oxidant and treatment of neuro degenerative disorders. There are many sources of quercetin as it is mostly found in fruits, vegetables and certain beverages. It is used in preventing injuries as they activate the anti-oxidant enzyme, reduction of alpha tocopheryl radicals, inhibits the oxidases, metal chelating activity, mitigation of oxidative stress caused by NO, increases uric acid level, increases the antioxidant properties of low molecular antioxidants and they also oxidize other compounds and act as pro oxidants.

Antioxidant properties:

Quercetin is a plant derived aglycone form of flavonoid glycosides. Quercetin is also supplied as a nutritional supplement as it is beneficial for many variety of diseases. Quercetin has many beneficial effects on cardiovascular protection, anticancer, antitumor, antiulcer, antiallergy, antiviral, anti-inflammatory activity, anti-diabetic, gastroprotective effects, antihypertensive, immunomodulatory, anti-infective, etc. It can also protect from the environmental causes of free radicals like smoking. Example of free radical is cigarette tar which damages the erythrocyte membranes.

Pharmacological importance:

Inflammation is a biological response of body that it shows during the harmful exposure of human body against harmful pathogen. Inflammation helps us in the healing process. Quercetin also have a property to modulate inflammation. It exhibits COX the inflammatory enzyme cyclooxygenase and lipoxygenase. The pharmacological importance of quercetin is shown up as anti-inflammatory, cardiovascular disease prevention, neurodegenerative disorders, antibacterial and anti-viral activity, allergies, asthma, hay fever and hives, and performs pharmacokinetics.

<u>Cancer and apoptosis:</u>

Many surveys collectively showed result that due to the consumption of more fruits and vegetables can help to reduce the risk of cancer. Quercetin have many properties such as growth factor suppression, antiproliferative that in anticancer properties. Flavonoid rich foods and vegetables helps us in cognition functions and also protect neurons by enhancing existing neuronal function. Also, flavonoids have stimulating neuronal regeneration factor.Quercetin is known for apoptosis inductor and has anticarcinogenic properties. Quercetin decreases the growth of tumor in liver, colon, brain and other tissues. It also inhibits the spread of malignant cells. Quercetin can reduce the number and size of ideal with minimal adverse effect. Both invitro and in vivo cancer studies proves that quercetin inhibits chemicals that are responsible and have a beneficial effect against prostate cancer.

<u>Pharmacokinetics:</u>

If quercetin is consumed in toxic dose, quercetin causes hypertension, reduction in serum potassium, emesis and nephrotoxicity. The distribution of quercetin is 0.7-708mm and elimination half life of intravenous quercetin is 3.8-86 min. Its volume of distribution is 3.7l/m square.

<u>Chemical effect of quercetin:</u>

Quercetin is proved to exert beneficial effects on health, including protection against lung cancer, cardiovascular and osteoporosis diseases. The third leading cause of death in the USA is by chronic obstructive pulmonary disease (COPD) which is reduced by the intake of flavonoids. This is the disease caused by disorder of the lung parenchyma and airways. Therapies for COPD, unfortunately said to be partially effect and may be beneficial in its treatment. In pre-clinical studies it has been demonstrated a 4-fold increase in the plasma quercetin level that helped to decrease lung inflammation and prevented disease progression

CUCURBITA:

Cucurbita are commonly known as pumpkin and belongs to the family known as Cucurbitaceae. This comes under cucurbit sp. Wild Cucurbita are too bitter to humans and livestock for feeding larger mammals and resistant to the bitter plants than the smaller ones. Pumpkin is mostly found everywhere but is specially belongs to North America. It is an annual creeping or climbing plant. It has a hollow center and is spherical in shape. It Has shallow root system that is well grown from taproot. Cucurbita has many remarkable properties as it is rich in nutrients, antibacterial, antidiabetic, antifungal, anticancer, antioxidant, anti-inflammatory, anti-ulcer. Nutrient content of pumpkin contains protein, magnesium, manganese, amino acid, fatty acid and zinc. Cucurbita is important to health to its abundance of vitamins and mineral content that our body needs and hence it is to be added in a good quantity in our daily diet. Cucurbita has unisexual flower and hence comes under monoecious plant with female and male on the same plant. Pumpkin flower produces nectar. Seeds of pumpkin is in network structure. Also, seeds are agro- industrial waste and is usually of no use. Seeds are off-white, flat, oval and flat disc in shape. Cucurbita can be consumed in any form either it be roasted, snack or cooked due to its taste of crispiness. Cucurbita has some phytochemical constituents such as flavonoids, alkaloids, oleic acid, palmitic and linoleic acid. Pumpkin provides herbal remedies as they are less toxic with limited side effects. Pumpkin can also be used as to derive oil, seed butter, chocolate, tortilla, candy, cereal bar, vegetable salad, cookies, chips, soups and muffins. It has beneficial effect on blood glucose, immunity, liver, bladder, prostate gland, cholesterol and hence is consumed on a large scale in many counties including India.

II. LITERATURE REVIEW

1) Neel Ratnam et al (2007) said Cucurbita pepo is widely used as food and herbal remedies around the world. It contains several phyto-constituents belonging to the categories of alkaloids, flavonoids, and palmitic, oleic and linoleic acids. Many pharmacological studies have demonstrated hepatoprotection, inhibit benign prostatic hyperplasia, hypoglycemic agent, antioxidant, anticancer, antimicrobial, anti-inflammatory, anti-diabetic, and antiulcer activities supporting its traditional uses. This aims a comprehensive study of the chemical constituents, pharmacological and clinical uses.

2) Karanja J.K et al (2013) explained the nutritional potential of thirteen varieties of Cucurbita fruits collected from selected regions of Kenya were evaluated for their proximate values, mineral and β-carotene content. The results indicated that the moisture content was high and it varied significantly ($p<0.05$) among the groups. Ash values were significantly different at ($p<0.05$) and ranged from 0.63 to 1.67%, crude protein was low in all the pumpkin groups and varied from 0.93 to 2.09%. The pumpkins also varied significantly ($p<0.05$) in crude fibre content (0.47- 1.95) among the groups. Potassium and sodium were the most prevalent minerals potentially useful amounts of potassium (199-172.3 mg/100g), sodium (56.39-118.28 mg/100g) and other essential minerals. The lowest micro-elements were copper and zinc. The results indicated that there were no significant differences among the groups in the β-carotene content ($p<0.05$). The nutrient information presented in this report should stimulate the nutritionists and agricultural officers in Kenya to consider the question of recommending the pumpkin to be consumed by adults and children alike in Kenya, including pregnant women and others with higher than normal nutritional requirements Keywords Nutritional, Cucurbita, Proximate, Mineral composition

3) Dr.P.Mythili Md et al (2017) suggested pumpkin (cucurbita maxima) belongs to cucurbitaceae which include pumpkin, gourds, melons and squashes. They are characterized by their seed oil. Pumpkin plant has been grown since the earliest history of mankind1. (Brucher – 1989). Pumpkin are grown around the world for variety of agricultural purposes, the seeds show anti helmintic, taenicide diuretic, cardio tonic, anti-inflammatory, anti - diabetic, antihyperlipidemic, anti – giardial activity CNS stimulant, melanogenesis inhibitory activity and immune modulatory action. Oil from seeds is nervine tonic. Cucurbita maxima seeds are good source of vitamins, minerals, calories and proteins. Multiple forms of phytase found in germinating cotyledons of cucurbita maxima   seeds. Seeds are rich in tocopherols (δ tocopherols), triterpene and unsaturated fatty acids.

4) Xiang D et al (2004) proposed Cucurbita maxima, belongs to family Cucurbitaceae, is commonly known as pumpkin. Several literature reports suggest it to be antidiabetic, antihypertensive, anticancer, immunomodulators, antibacterial and antihyperlipidaemic. One important approach for the treatment of type 2 diabetes mellitus is by decreasing the postprandial hyperglycemia in effect. This is possible by inhibiting certain carbohydrate hydrolyzing enzyme like α-amylase and α-glucosidase. The objective of present study was to evaluate in-vitro antioxidant and α-amylase inhibitory activity of methanolic extract of C. maxima leaves. Methanolic extract of leaves of C. maxima showed strong antioxidant (DPPH scavenging) and α-amylase inhibition (IC50 125 µg/ml and IC50 2.1 mg/ml), respectively. These results suggest the possible use of pumpkin leaves in the management of diabetes mellitus.

5) Jacks TJ et al (1972) investigated quercetin and allopurinol significantly inhibited the TXNIP overexpression, activation of NLRP3 inflammasome, down-regulation of PPARα and up-regulation of sterol regulatory element binding protein-1c (SREBP-1c), SREBP-2, fatty acid synthase and liver X receptor α, as well as elevation of ROS and IL-1β in diabetic rat liver. These effects were confirmed in hepatocytes in vitro and it was further shown that TXNIP down-regulation contributed to the suppression of NLRP3 inflammasome activation, inflammation and changes in PPARα and SREBPs.

6) Ang-Lee MK et al (2001) studied dietary plants and herbal preparations have been traditionally used as medicine in developing countries and obtained a resurgence of use in the United States and Europe. Research carried out in last few decades has validated several such claims of use of traditional medicine plants. Popularity of pumpkin in various systems of traditional medicine for several ailments (antidiabetic, antihypertensive, antitumor, immunomodulation, antibacterial, antihypercholesterolemia, intestinal antiparasitia, antiinflammation, antalgic) focused the investigators' attention on this plant. Considerable evidence from several epidemiological studies concerning bioactivities leads have stimulated a number of animal model, cell culture studies and clinical trials designed to test this pharmacological actions. In addition, it was found that technologies such as germination and fermentation could reduce antinutritional materials and affect the pharmacological activities of pumpkin. This review will focus on the main medicinal properties and technologies of pumpkin, and point out areas for future research to further elucidate mechanisms whereby this compound may reduce disease risk.

7) Decker-Walters DS et al (2000) says a review of the activities of members of the Cucurbitaceae was carried out. Many of them are confirmed nutritious and therapeutical. Their global spread, diverse genera and phytochemical profile further confirm them as an attraction for the growth and survival of humanity. The need for alternative control measures to address resistance has heightened the passion for Cucurbitaceae in bioprospecting.

8) M.Manasa et al (2001) explained herbal medicines have gained global importance with bothmedicinal and economic implications these days. The scientific assessment has become a prerequisite for acceptance of health claims. Onion (Allium cepa) a member of genus Allium, is the second most important horticulture crop all over the world. It is used as important source of phytoconstituents and food flavor. Allium cepa is one of the richest source of flavonoids and organosulphur compounds.Wide spectrum of biological activities makes A.cepa a potential therapeutic agent.

9) Cai X et al (2013) suggested quercetin (QC) is a typical plant flavonoid, possesses diverse pharmacologic effects including antiinflammatory, antioxidant, anti-cancer, anti-anaphylaxis effects and against aging. However, the application of QC in pharmaceutical field is limited due to its poor solubility, low bioavailability, poor permeability and instability. To improve the bioavailability of QC, numerous approaches have been undertaken, involving the use of promising drug delivery systems such as inclusion complexes, liposomes, nanoparticles or micelles, which appear to provide higher solubility and bioavailability. Enhanced bioavailability of QC in the near future is likely to bring this product to the forefront of therapeutic agents for treatment of human disease.

10) Ravi Kant Upadhyay et al (2016) investigated cucurbita contain phenolics and flavonoids that have potential anti-inflammatory, anti-cholesterol, anticancer, and antioxidant properties. Onions contain 89% water, 1.5% protein, and vitamins B1, B2, and C, along with potassium and selenium. It also contains polysaccharides such as fructosans, saccharose, peptides, flavonoids (mostly quercetin), and essential oil. Onion contains numerous sulfur compounds including thiosulfinates and thiosulfonates; cepaenes; S-oxides; S, S-dioxides; mono, di, and tri-sulfides; and sulfoxides. Onion is highly nutritional and its dietary use improves digestion and mental health and lower down toxigenicity of oils. Onion has potential in treating cardiovascular disease, hyperglycemia, and stomach cancer. Onion contains an important antioxidative, i.e., quercetin that is derived from Allium cepa on aldehyde oxidase low-density lipoprotein which reduces hepatocytes apoptosis in streptozotocin-induced diabetic rat. Onion has great ethnomedicinal importance as native remedies used against diabetes, and related complications are from onion. A. cepa red and white varieties showed antimicrobial and antioxidant activities. These are used in traditional Indian spices and are of great health significance. These are curative for implications from and for food cultures for cardiovascular disease and provide longevity

11) Manasa Machavarapu et al (2013) noted allium cepa is the most common and richest natural source of flavonoids. In this study the extraction of flavonoids quercetin and rutin and the total phenolic compounds were carried out. Optimization of physico-chemical parameters namely effects of different solvents, different solvent percentages, soaking time, different amounts of sample (Skin of Allium cepa) and different pH were studied. For the extraction of quercetin, the optimum results were 80% (v/v), 1d, 1% (w/v) and pH 5.5 respectively. The highest quercetin and rutin concentration for optimized conditions were 185.309 µg/L and 124.69 µg/L respectively and it was found to be 19.412 mg/L for total phenolic content

12) Bhimanagouda S. Patil et al (1995) studied the aglycone, or free quercetin, and total quercetin content of 75 cultivars and selections was analyzed by reverse-phase high-performance liquid chromatography. Quercetin glycosides were hydrolyzed into aglycones. Total quercetin content in yellow, pink, and red onions varied from 54 to 286 mg·kg $^{-1}$ fresh weight in different onion entries grown during 1992. White onions contained trace amounts of total quercetin. Free quercetin content in all the onions was low (< 0.4 mg·kg $^{-1}$) except in `20272-G' (12.5 mg·kg $^{-1}$ fresh weight). Bulbs stored at 5, 24, and 30C and controlled atmosphere (CA) for 0,1,2,3,4, and 5 months showed a most marked change in total quercetin content at 24C compared to other treatments, with a rise in mid-storage followed by a drop. Storage at 5 and 30C also demonstrated a similar change. However, total quercetin content did not vary significantly in bulbs stored at CA for 5 months. We conclude that genetic and storage factors affect quercetin content on onions

13) Bansode Ashwini Sopan et al (2014) proposed as Antioxidant potential of many plants is mainly due to phenolic components such as phenolic acid, phenolic diterpense therefore total phenolic content and antioxidant potential of cucurbita maxima. (Pumpkin) powder was determined respectively by using Folin-Ciocalteu reagent and diphenyl-picryl-hydrazyl (DPPH) assay. In this method the antioxidants present in the plant extracts reacted with DPPH, which is a stable free radical and converted it to 1,1-diphenyl-1,2-picryl, hydrazine. Total phenolic content of pumpkin was determined using Folin-Ciocalteu reagent and gallic acid as standard, amount of phenolic compound was found to be 6.5 mg/gm powder of the drug. The high fiber content of pumpkin helps to aid proper digestion and in traditional medicine it is used as an anthelmintic, taenicide, diuretics and in the treatment of hemorrhoids so we compared the anti oxidant potential of pumpkin powder with marketed preparation (Pilex tablet). The half maximal inhibitory concentration (IC50) is a measure of the effectiveness of a compound in inhibiting biological or biochemical function. IC50 value of pilex tablet = 8.53 µg/ml & IC50 value of pumpkin powder =9.12 µg/ml.

14) Md. Irshad et al (2014) noted pumpkin, as a dietary plant, has been used in traditional medicine around the world. In addition, during the last decade, antidiabetic, antihypertensive, antitumor, intestinal antiparasitic, antibacterial, anti hypercholesterolemia, anti-inflammatory, immunomodulatory and analgic effects of pumpkin has been reported. The aim of the present study was to determine the effects of different extracts of Cucurbita pepo L. on the immune responses. Methanolic, chloroform and ethylacetate extracts of C. pepo fruits was obtained using percolation method. Mice were used to study the effects of C. pepo extracts on the acquired immunity. Sheep red blood cell (SRBC) was injected (S.C., 1×108 cells/ml, 20 µl) and 5 days later, methanolic, chloroform and ethylacetate extracts of C. pepo at diiferent doses (10, 100 and 500 mg/kg), betamethasone and levamisol at equal doses (4 mg/kg) as positive controls and normal saline as a negative control were given i.p. After 1 h SRBC was injected to the footpad (S.C., 1×108 cells/ml, 20 ml) and the footpad swelling was measured up to 72 h. To investigate the effects of C. pepo on the innate immunity the same procedure was used, but animals received only one injection of SRBC 1 h after i.p. injection of test compounds. Our findings showed that SRBC induced an increase in the paw swelling with maximum response at 6-8 h. Betamethasone inhibited the paw swelling in both models. In both innate and acquired immunity models, methanolic, chloroform and ethylacetate extracts of C. pepo fruits significantly reduced the paw swelling dose dependently. The data suggest that the pumpkin extracts may have immunomodulatory effects.

15) Mortada M. El-Sayed et al (2016) says this study aimed to determine total phenolic and flavonoid contents as well as evaluate antioxidant activity of the defatted methanolic extracts of flowers, leaves, stems and certain fractions (dichloromethane, ethyl acetate, n-butanol and water residue) derived from the defatted methanolic extract of Lantana camara and Cucurbita pepo (Squash) leaves besides, the essential oils of L. camara leaves and flowers by using (DPPH, ABTS and TAC assay). The chemical composition of L. camara essential oils was identified by GC-MS analysis. The results exhibited that the defatted methanolic extracts of L. camara (flowers, leaves and stems) have higher phenolic and flavonoid contents as well as antioxidant activity than C. Pepo (Squash) parts. Also, the ethyl acetate fraction of L. camara has the highest phenolic and flavonoid contents as well as the highest antioxidant activity (DPPH; SC 50 = 11.82±0.21 µg/ mL, TAC; 290.96±6.71mg eq. ascorbic acid /g extract, ABTS; 94.03±0.25mm Trolox ® eq. / 100g extract). There is a positive correlation between the total phenolic and flavonoid contents and the antioxidant activity of the tested extracts or fractions. GC-MS analysis of essential oil of L. camara leaves and flowers revealed the presence of 47 compounds in leaves and 40 compounds in flower. The major compounds in the leaf oil were 7(11)-selinen-4α-ol (14.50%) and cedrenol

16) Shibam Mondal et al (2017) explainted a novel antifungal peptide, with a molecular mass of 8 kDa in sodium dodecyl sulfate-polyacrylamide gel electrophoresis and in gel filtration on Superdex 75 and designated cucurmoschin, was isolated from the seeds of the black pumpkin. The peptide was unadsorbed on DEAE-cellulose but adsorbed on Affi-gel blue gel.

Cucurmoschin inhibited mycelial growth in the fungi Botrytis cinerea, Fusarium oxysporum and Mycosphaerella oxysporum. It inhibited translation in a cell-free rabbit reticulocyte lysate system with an IC50 of 1.2 microM. The N-terminal sequence of cucurmoschin was rich in arginine, glutamate and glycine residues.

17) Seun F. Akomolafe et al (2016) suggested pumpkin seed has been associated with myriad of medicinal uses in different part of the world. In this study, phenolic composition and Fe2+ induced thiobarbituric acid reactive species (TBARS) inhibitory ability of methanolic extract from pumpkin seeds in rat's testes homogenates were determined. The extract was prepared with 80% methanol (v/v) and the radicals [(1,1-diphenyl–2 picrylhydrazyl (DPPH), 2,2'-azino-bis(3-ethylbenzthiazoline-6-sulphonic acid) (ABTS)] scavenging, Fe2+ chelation and ferric reducing abilities of the extract were carried out. The phenolics composition was also investigated using gas chromatography couple with flame ionization detector (GC-FID). The GC analysis revealed the presence of vallinic, coumaric protocatechuic, caffeic, ferulic and sinapinic acids, and apigenin, quercetin, luteolin, kaempferol as the dominant phenolic compounds. The results revealed that the extract inhibited Fe2+-induced TBARS, scavenge DPPH radical and chelate Fe2+ in a dose dependent manner. The extract also scavenged ABTS radical and reduced Fe2+ to Fe3+. Although, the standard used had higher effect compared to the extract, nevertheless, the TBARs inhibitory potential of the extracts clearly gives an insight on the protective potentials against oxidative induce testicular damage that might lead to male infertility if unchecked. These abilities could however be linked to the presence of polyphenolic compounds.

18) Jyotsana Singh et al (2016) investigated cucurbits are major economically important species of plants; particularly those with edible fruits having nutritional significance. The present work was to investigate the polyphenolic content and antioxidant capacity of peels and pulps of four cucurbit fruits, namely pumpkin, ash gourd, watermelon and muskmelon. The solvent systems used were methanol, ethanol and acetone at three different concentrations in distilled water (50, 70, and 100%) and 100% distilled water. The extracts were analyzed for their total phenolic content, total flavonoid content and antioxidant activities using Ferric Reducing Antioxidant Power assay (FRAP assay), DPPH free radical-scavenging assay and ABTS radical scavenging capacity. The result showed that the highest extraction was by 50% acetone in case of peels and

50% ethanol in case of pulp. The best solvent was 50% acetone as it gave highest yield as well as showed highest correlation between various assays. The polyphenolic content and the antioxidant activity were high in peels than pulps. The muskmelon fruit extracts (peel and pulp) showed highest antioxidant activity. High polyphenolic content showed significant correlation with high antioxidant activity. The result indicated that these cucurbit fruit are good source of natural antioxidants which can be utilized as an ingredient to functional foods.

19) Suman Chandra et al (2014) studied pumpkin solid wastes, composed of the apical trimmings and the outer dry layers, were used as raw material for the recovery of antioxidant phenolics by employing acidified water/ethanol-based solvent systems. The extraction efficiency was assessed by measuring the reducing power and the antiradical activity of the extracts, and the total polyphenol yield. Improvement of the extraction yield was based on modifying the solvent composition and the operating conditions (temperature, time). Initial screening of various mixtures showed that significantly higher recoveries ($P < 0.001$) can be attained by using 60% ethanol containing 0.1% HCl. The assessment of factors affecting yield, including extraction time and temperature, was accomplished using a series of extractions on the basis of a 2×3 factorial design model, with time and temperature values varying from 0.5 to 6h and from 40 to 60°C, respectively. The results obtained showed that improvement of yield is dependent upon increasing extraction time up to 6 hours, whereas increasing the temperature from 40 to 60°C had a negative effect.

20) Jeong-Yeon Ko et al (2016) proposed enhanced production of individual phenolic compounds by subcritical water hydrolysis (SWH) of pumpkin leaves was investigated at various temperatures ranging from 100 to 220°C at 20 min and at various reaction times ranging from 10 to 50 min at 160°C. Caffeic acid, p-coumaric acid, ferulic acid, and gentisic acid were the major phenolic compounds in the hydrolysate of pumpkin leaves. All phenolic compounds except gentisic acid showed the highest yield at 160°C, but gentisic acid showed the highest yield at 180°C. The cumulative amount of individual phenolic compounds gradually increased by 48.1, 52.2, and 78.4 µg/g dry matter at 100°C, 120°C, and 140°C, respectively, and then greatly increased by 1,477.1 µg/g dry matter at 160°C. The yields of caffeic acid and ferulic acid showed peaks at 20 min, while those of cinnamic acid, p-coumaric acid, p-hydroxybenzoic acid, and procatechuic acid showed peaks at 30 min. Antioxidant activities such as 2,2-diphenyl-1-picrylhydrazyl and ferric reducing antioxidant power values gradually increased with hydrolysis temperature and ranged from 6.77 to 12.42 mg ascorbic acid equivalents/g dry matter and from 4.25 to 8.92 mmol Fe2+/100 g dry matter, respectively. Color L* and b* values gradually decreased as hydrolysis temperature increased from 100°C to 140°C. At high temperatures (160°C to 220°C), L* and b* values decreased suddenly. The a* value peaked at 160°C and then decreased as temperature increased from 160°C to 220°C. These results suggest that SWH of pumpkin leaves was strongly influenced by hydrolysis temperature and may enhanced the production of phenolic compounds and antioxidant activities.

21) Echessa A.C.Peter et al (2013) noted free radical induced oxidative stress is involved in the pathogenesis of various diseases and disorders. Antioxidants play an important role against this oxidative stress to protect our body. The present study was carried out to evaluate the in vitro antioxidant properties of methanol extract of C.maxima aerial parts (MECM). Methods: MECM was assayed on different in vitro free radical models like, DPPH, nitric oxide, superoxide, hydrogen peroxide and lipid peroxide radical models. Reductive ability of the extract was also tested by the complex formation with potassium ferricyanide. Further total phenolic and flavonoid contents of the crude extract were also measured. Butylated hydroxy toluene was taken as standard.

22) Roza Martha Perez Gutierrez et al (2016) says doxorubicin-based chemotherapy induces cardiotoxicity, which limits its clinical application. We previously reported the protective effects of quercetin against doxorubicin-induced hepatotoxicity. In this study, we tested the effects of quercetin on the expression of Bmi-1, a protein regulating mitochondrial function and ROS generation, as a mechanism underlying quercetin-mediated protection against doxorubicin-induced cardiotoxicity.

23) Dhiman K et al (2012) explained n H9c2 cells, quercetin reduced doxorubicin-induced apoptosis, mitochondrial dysfunction, ROS generation and DNA double-strand breaks. The quercetin-mediated protection against doxorubicin toxicity was characterized by decreased expression of Bid, p53 and oxidase (p47 and Nox1) and by increased expression of Bcl-2 and

Bmi-1. Bmi-1 siRNA abolished the protective effect of quercetin against doxorubicin-induced toxicity in H9c2 cells. Furthermore, quercetin protected mice from doxorubicin-induced cardiac dysfunction that was accompanied by reduced ROS levels and lipid peroxidation, but enhanced the expression of Bmi-1 and anti-oxidative superoxide dismutase.

24)) Dubey A et al (2010) suggested the food is the medicine" this statement led us to explore the antimicrobial and antioxidant potential of Cucurbita pepo L. The plant belongs to family Cucurbitaceae. Cucurbits have a lot of folkloric medicinal uses. In this study, in vitro spectroscopic protocols were executed for the determination of total phenolic content, the concentration of flavonoids, total antioxidant capacity as well as free radical scavenging potential of leaves extract of Cucurbita pepo L. (CPL) in solvents i.e., n-hexane, chloroform, ethyl acetate, n-butanol, ethanol, and water. Ethyl acetate extract of CPL. exhibited the highest flavonoid concentration, whereas n-Butanolic extract was found to contain highest total phenolic content and both extracts also displayed a strong antioxidant activity. The antioxidant activity of extracts was expressed as percentage of DPPH radical inhibition and IC50 values (µg/ml). A significant linear correlation was observed between the values for the total phenolic content, flavonoid concentration, phosphomolybdate assay and percentage antioxidant activity of all the extracts. The higher concentration of phenolic compounds was inferred to be a major contributor of antioxidant activity. Antimicrobial potential of various extracts of CPL leaves were explored. Aqueous, n-Butanolic and ethyl acetate extracts were found active against Bacillus subtilis, Pseudomonas aeruginosa, and Staphylococcus aureus. Therefore, Cucurbita pepo L. CPL leaves can be considered as an auspicious candidate for natural plant source of especially antioxidants and antimicrobial agents.

25) Sun J et al (2005) invested molecular weight, degree of esterification, monosaccharide composition and gelation behaviour of different samples of pectin from sugar beet and pumpkin have been investigated. Monosaccharide composition and degree of esterification were determined simultaneously by gas—liquid chromatography. Each of the samples studied contained glucose, galactose, rhamnose, arabinose and galacturonic acid in different proportions.

26) Saha P et al (2011) proposed cucurbitacins which are structurally diverse triterpenes found in the members of Cucurbitaceae and several other plant families possess immense pharmacological potential. This diverse group of compounds may prove to be important lead molecules for future research. Research focused on these unattended medicinal leads from the nature can prove to be of immense significance in generating scientifically validated data with regard to their efficacy and possible role in various diseases. This review is aimed to provide an insight into the chemical nature and medicinal potential of these compounds exploring their proposed mode of action, probable molecular targets and to have an outlook on future directions of their use as medicinal agents.

27) C. Madhavi Latha et al (2011) studied flavonol and flavone glycosides occur in common vegetables, mainly as the quercetin or kaempferol glycosides and less frequently as the luteolin or apigenin glycosides. Their formation normally depends on light, so they are mainly concentrated in the outer tissues. Their concentration in free-standing leaves is considerably greater than that in any other part of the same plant, with the exception of the onion.

28)Xanthopoulou MN et al (2009) noted pumpkin seeds have been implicated in providing health benefits. However their antioxidant or anti-inflammatory activity of their extracts has never been studied. Therefore, four commercially available pumpkin seeds were treated with two different extraction methodologies in order to obtain fractions with different content. The extracts were screened for their antioxidant activity using 2,2-diphenyl-1-picrylhydrazyl (DPPH) free radical scavenging assay and for their inhibitory activity against lipid peroxidation catalyzed by soybean lipoxygenase. Most extracts tested have demonstrated radical scavenging activity, which depends on their total phenolic content, with fractions rich in phenolics showing the strongest activity. On the other hand, the phenolic content of extracts does not determine their activity against lipoxygenase, as acetone and polar lipid fractions are its strongest inhibitors. The presence of molecules being able to scavenge radicals and inhibit lipoxygenase in pumpkin seeds may in part explain the health benefits attributed to them.

29) Craig WJ et al (1997) says consuming a diet rich in plant foods will provide a milieu of phytochemicals, nonnutritive substances in plants that possess health-protective benefits. Nuts, whole grains, fruits, and vegetables contain an abundance of phenolic compounds, terpenoids, pigments, and other natural antioxidants that have been associated with protection from and/or treatment of chronic disease such as heart disease, cancer, diabetes, and hypertension as well as other medical conditions. The foods and herbs with the highest anticancer activity include garlic, soybeans, cabbage, ginger, licorice, and the umbelliferous vegetables. Citrus, in addition to providing an ample supply of vitamin C, folic acid, potassium, and pectin, contains a host of active phytochemicals. The phytochemicals in grains reduce the risk of cardiovascular disease and cancer.

30) Nair R et al (2005) explained there is significant evidence that the pathogenesis of several neurodegenerative diseases, including Parkinson's disease, Alzheimer's disease, Friedreich's ataxia and amyotrophic lateral sclerosis, may involve the generation of reactive oxygen species and mitochondrial dysfunction. Here, we review the evidence for a disturbance of glutathione homeostasis that may either lead to or result from oxidative stress in neurodegenerative disorders. Glutathione is an important intracellular antioxidant that protects against a variety of different antioxidant species. An important role for glutathione was proposed for the pathogenesis of Parkinson's disease, because a decrease in total glutathione concentrations in the substantia nigra has been observed in preclinical stages, at a time at which other biochemical changes are not yet detectable. Because glutathione does not cross the blood–brain barrier other treatment options to increase brain concentrations of glutathione including glutathione analogs, mimetics or precursors are discussed.

III.EXPERIMENTAL PROCEDURE

MATERIALS AND METHODS

I.Chemicals and reagents:

Ethanol, Aluminium chloride, Potassium acetate.

II.Collection of plant material:

Peel of *Cucurbita* is collected from a local market, Meerut area. These are cleaned and dried. The dried seeds are powdered and used as a raw.

Peel of Cucurbita

Powder form peel

I. Soxhlet extraction process

200ml of ethanol is taken at round bottom flask. 10grams of peel of *cucurbita* powder was placed in thimble and fix it to the condenser. Now the total apparatus were placed in the heater. Using the soxhlet apparatus continuous extraction was done for 4 hrs and heat the filtrate solution at 78^0c [4]. So that the solvent which is taken in the glass wear is evaporated and finally stored that extract sample.

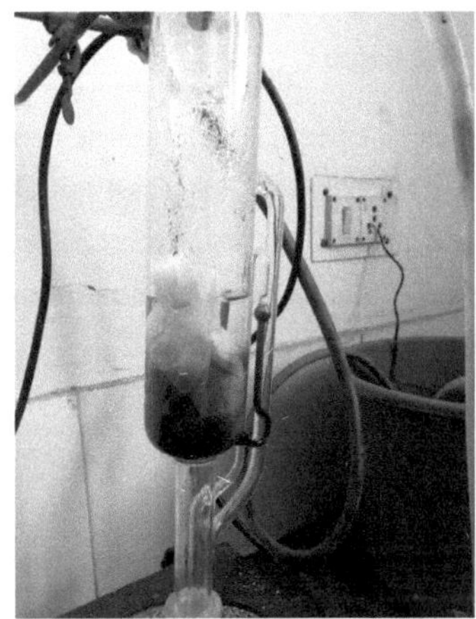

Soxhlet Extraction Process

II. Fermentation Process

Microbial growth:

100ml of Nutrient broth medium is autoclaved at 121°C for 15 min and distributed the nutrient broth into 250ml conical flasks. Bacillus cereus was taken from culture medium and inoculated into above conical flasks in the vicinity of laminar air flow. Incubate the cultures in the orbital shaker incubator at 30 °C temperature at 120 rpm for 24hr [3].

Fermented Extract

Take 50 ml of ethanolic extract sample from soxhlet extraction process. 20ml of 5% of glucose solution (carbon source) was added to extract. *Bacillus cereus* (1ml) was taken from culture medium and inoculated into extract of peel of *cucurbita*.Then flask was incubated at room temperature. After the specified incubation period, the fermented samples were filtered through Whatman No.1 filter paper. The filtrate was centrifuged at 10,000 rpm for 10 min at 4 °C. After centrifugation, collect the supernatant as fermented sample. In the fermented sample, quercetin and rutin were determine by using aluminum chloride spectrophotometer method [1,5].

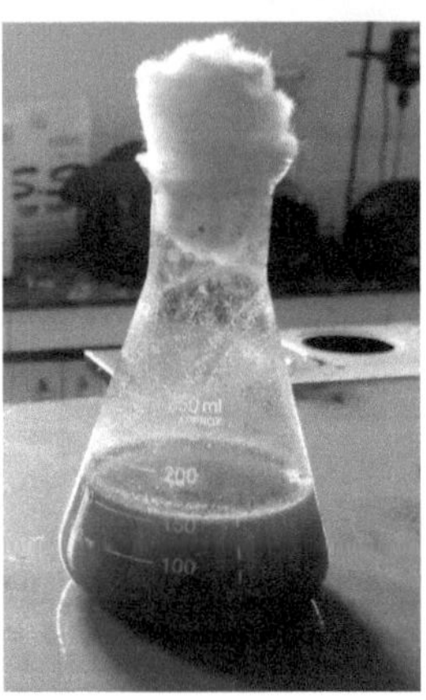

Fermentation sample

Estimation of Quercetin

0.5ml of ethanolic extract taken in a test tube and 1.5 ml of methanol, 0.1 ml of 10% aluminium chloride, 0.1 ml of 1M potassium acetate and 2.8 ml of distilled water were added. The mixture was allowed to stand for 30 min at room temperature. The absorance of the reaction mixture was measured at 415 nm using colorimeter. The Quercetin was determined by using calibration curve [6].

Estimation of Rutin

Each 0.5ml of water extracts and fermented sample taken in each test tube and add 0.5 ml of 2% aluminium chloride ($AlCl_3$) solution dissolved in ethanol. The mixture was allowed to stand for 30 min at room temperature. The absorbance of the reaction mixture was measured at 420 nm using spectrophotometer. The rutin was determined by using calibration curve [6,7].

III. Anti cancer activity

After fermentation process, the fermented or incubated sample is used for anticancer activity by using MTT assay method. Here anticancer cells are Caco-2 colon cancer cell lines.

Cell culture

Human cancer cell lines used in this study were procured from National Centre for Cell Science, Pune. All cells were grown in Minimal essential medium (MEM) supplemented with 4.5 g/L glucose, 2 mM L-glutamine and 5% fetal bovine serum (FBS) (growth medium) at 37°C in 5% CO_2 incubator.

Flow chart of MTT assay

Trypsinized cells

⇩

Cells are seeded in to 96-well plate

⇩

Add supplements of growth medium & cultured at 37°C in 5%CO_2

⇩

After 48hr incubation, centrifugation (10,000 rpm &4 °C), supernatant was discared

⇩

Again the cells (pellet) were pretreated with growth medium

⇩

Subsequently mixed with different concentrations of test compounds (6.25 to 200 µg/ml)

⇩

Cultured for 48hr

⇩

Each well, 20µl of fresh MTT & incubation 2hr at 37°C & colored formazon product was formed

⇩

Supernatant growth was removed from each well

⇩

Add 10µl of DMSO to solubilize formazon produ product

⇩

After 30 minutes incubation the absorbance of the culture plate was read at 570nm ELISA reader

IV. RESULTS

Extraction of Quercetin from peel of *Cucurbita* by Soxhlet Extractor

In the soxhlet extraction process, 10 gm of peel of *cucurbita* powder is placed in a thimble, and 200ml of ethanolic solvent is placed in the round bottom flask [8]. The water droplets are coming out from the condenser and fall on the powder. In every cycle, only flavor color is coming out from the powder and falls into the ethanolic solvent. At 4 hr it showed maximum of 18.0µg/ml.

Enhancement of Quercetin by Fermentation Process

In the present work, 18.0µg/ml of quercetin was obtained from soxhlet extraction process. Now this extract sample was as fermented sample. In the fermentation process, the quercetin concentration was enhanced and it is found to be 37.0µg/ml. The quercetin concentration is increased from fermentation extract due the secretion of β-glucosidase enzyme [1,9] which may converts rutin to quercetin. This enzyme was produced by *Bacillus cereus* which converts rutin to form quercetin. So this process of biotransformation of rutin to quercetin carried out by *Bacillus cereus*.

Table: Enhancement of Quercetin

S.No	Time (hr)	Quercetin (µg/ml)	Rutin (µg/ml)
1	5	20.5	36.0
2	10	21.0	35.0
3	15	22.5	33.5
4	20	24.5	32.0
5	25	25.5	31.0
6	30	27.0	29.0

7	35	27.5	28.5
8	40	29.5	27.5
9	45	30.5	26.0
10	50	32.5	25.0
11	55	33.5	24.5
12	60	35.0	23.0
13	65	35.5	22.5
14	70	36.5	21.5
15	75	37.0	19.5
16	80	37.0	19.5
17	85	37.0	19.5

Anti cancer

The percentage inhibition of cancer cell lines can be calculated as

$100 - (At-Ab) / (Ac-Ab) \times 100$

At: Absorbance of test,

Ab: Absorbance of blank,

Ac: Absorbance of control.

Dose Response of incubated sample on Caco2 (Colon Cancer) Cell line

Blank   :   0.040

Control   :   0.468

Conc (ug/ml)	OD of 5' fluorouracil (STD)at 570 nm	% Cell Survival	% Cell Inhibition	OD of incubatede d sample	% Cell Survival	% Cell Inhibition
6.25	0.407	82.3	17.7	0.442	98.5	1.5
12.5	0.243	45.5	54.5	0.417	92.4	7.6
25	0.200	37.4	62.6	0.386	84.8	15.2
50	0.165	29.2	70.8	0.322	69.1	30.9
100	0.130	21	79	0.268	60.3	39.7
200	0.088	11.2	88.8	0.204	40.2	59.8

Dose Response of extract on Caco-2 (Colon Cancer) Cell line

Conc of incubate sample (ug/ml)	% Cell Inhibition
6.25	1.5
12.5	7.6
25	15.2

50	30.9
100	39.7
200	59.8
IC$_{50}$	**150.86** **µg/mL**

This plant extract induces a cell arrest to inhibit the growth of the Caco2 colon cancer cells from 1.5% to 59.8%.

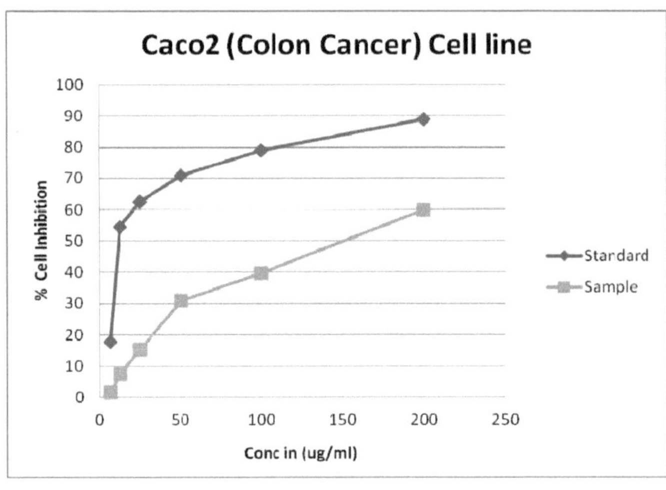

DISCUSSION

In the soxhlet extraction process, 10 gm of peel of *cucurbita* powder is placed in a thimble, and 200ml of ethanolic solvent is placed in the round bottom flask. The water droplets are coming out from the condenser and fall on the powder. In every cycle, only flavor color is coming out from the powder and falls into the ethanolic solvent [10]. At 4 hr it showed maximum of 18.0µg/ml. In the present work, 18.0µg/ml of quercetin was obtained from soxhlet extraction process. Now this extract sample was as fermented sample. In the fermentation process, the quercetin concentration was enhanced and it is found to be 37.0µg/ml. The quercetin concentration is increased from fermentation extract due the secretion of β-glucosidase enzyme [1] which may converts rutin to quercetin. This enzyme was produced by *Bacillus cereus* which converts rutin to form quercetin. So this process of biotransformation of rutin to quercetin carried out by *Bacillus cereus*. This fermented sample or incubated sample is used for anticancer studies. This fermented sample induces a cell arrest to inhibit the growth of the Caco2 colon cancer cells from 1.5% to 59.8%.

V. CONCLUSION

The quercetin production was found to best results from fermentation process than extraction process due to biotransformation. In this research, the highest production of quercetin was found to be 18.0µg/ml and 37.0µg/ml from the extraction process and the fermentation process. The fermented sample or incubated sample is used for anticancer studies. This fermented sample induces a cell arrest to inhibit the growth of the Caco2 colon cancer cells from 1.5% to 59.8%.

VI. REFERENCES

1)Neel Ratnam, Vandana, Md Najibullah, MD Ibrahim (2017). A review on cucurbita pepo. International Journal of Pharmacognosy and Phytochemicals research, 9(9): 1190-1194.

2)Karanja J.K, Mungindi B.J, Khamis F.M, Muchugi A.N (2013). Nutritional Composition of the Pumpkin (cucurbita species) seeds cultivation from selected regions in Kenya.Journal of Horticulture Letters, 3(1): 17-22.

3) Dr.P.Mythili Md, Dr.T.Kavitha Md (2017). Overview on cucurbita Maxima seed. IOSR Journal of Dental and Medical sciences (IOSR-JDMS),16(3): 29-33.

4)Xiang D, Han FY, Liang P (2004) Extraction of pumpkin polysaccharide with sodium hydroxide. Sci Technol Food Ind 11: 120–122.

5)Jacks TJ, Hensarling TP, Yatsu LY (1972) Cucurbit seeds: I. Characterizations and Uses of oils and Proteins. A Rev Econ Bot 26: 135–141.

6)Ang-Lee MK, Moss J, Yuan CS (2001) Herbal medicines and perioperative care. JAMA 286: 208–216.

7)Decker-Walters DS, Walters TW (2000) Squash. In: Kipel KF, Ornelas KC (eds), The Cambridge World History of Food. Cambridge, England: Cambridge Univ. Press, pp 335–351

8)M.Manasa, S.Manoj Kumar and Meena Vangalapati. A review on medicinal herb : Allium

9)Cai X, Fang Z, Dou J, Yu A, Zhai G ,Bioavailability of quercetin: problems and promises , Journal of Current Medicinal Chemistry 2013;20(20):2572-82.

10)Ravi Kant Upadhyay. Nutraceutical, pharmaceutical and therapeutic uses of Allium cepa: A review. International Journal of Green Pharmacy. 2016; 10 (1) : 46-64.

11)Manasa Machavarapu, Manoj Kumar Sindiri and Meena Vangalapati. Optimization of physic-chemical parameters for the extraction of flavonoids and phenolic compounds from the skin of Allium cepa. 2013, 2(7): 3125-3129.

12)Bhimanagouda S. Patil, Leonard M. Pike, and Kil Sun Yoo. Variation in the Quercetin

Content in Different Colored Onions (Allium cepa L.), Journal of American Society for Horticultural Science, 120(6):909-913. 1995.

13)Bansode Ashwini Sopan, Devhadrao Nitin Vasantrao, S. Bansode Ajit. Total Phenolic Content and Antioxidant Potential of Cucurbita Maxima (Pumpkin) Powder, International Journal of Pharmaceutical Sciences and Research, 2014; 5(5): 1903-1907p.

14) Md. Irshad, Irfan Ahmad, Syed Jafar Mehdi , Harish Chandra Goel, M. Moshahid A. Rizvi. Antioxidant Capacity and Phenolic Content of the Aqueous Extract of Commonly Consumed Cucurbits, International Journal of Food Properties, 2014; 17:179–186.

15)Mortada M. El-Sayed, Maher M. Hashash, Afaf A. Abdel-Hady, Heba Abdel-Hady, Ezzat E. Abdel-Lateef, Eman A.Morsi. Total Phenolic and Flavonoid Contents and antioxidant activity of Lantana Camara and Cucurbita Pepo (Squash) Extracts as well as Gc-Ms analysis of Lantana Camara essential oils, World Journal of Pharmaceutical Research. 2016; 6(1): 137-153.

16) Shibam Mondal, Imrul Hossain, Md. Nur Islam. Determination of antioxidant potential of Cucurbita pepo Linn. (An edible herbs of Bangladesh), Journal of Pharmacognosy and Phytochemistry. 2017; 6(5): 1016-1019.

17) Seun F. Akomolafe, Ganiyu Oboh, Sunday I. Oyeleye, Olorunfemi R. Molehin, Opeyemi B. Ogunsuyi. Phenolic Composition and Inhibitory Ability of Methanolic Extract from Pumpkin (Cucurbita pepo L) Seeds on Fe-induced Thiobarbituric acid reactive species in Albino Rat's Testicular Tissue In-Vitro, Journal of Applied Pharmaceutical Science, 2016; 6(9):115-120.

18) Jyotsana Singh, Vinti Singh, Surabhi Shukla, Awadhesh K Rai. Phenolic Content and Antioxidant Capacity of Selected Cucurbit Fruits Extracted with Different Solvents, Journal of Nutrition & Food Sciences. 2016; 6(6): 1-8.

19) Suman Chandra, Shabana Khan, Bharathi Avula, Hemant Lata, Min Hye Yang, Mahmoud A. ElSohly, Ikhlas A. Khan. Assessment of Total Phenolic and Flavonoid Content, Antioxidant Properties, and Yield of Aeroponically and Conventionally Grown Leafy Vegetables and Fruit Crops: A Comparative Study, Journal of Evidence-Based Complementary and Alternative Medicine, 2014; 1(1):1-9.

20)Jeong-Yeon Ko, Mi-Ok Ko, Dong-Shin Kim, Sang-Bin Lim. Enhanced Production of Phenolic Compounds from Pumpkin Leaves by Subcritical Water Hydrolysis, Preventive Nutrition and Food Sciences Journal, 2016; 21(2):132-137.

21)Echessa A.C.Peter, Nyambaka Hudson, Ondigi N.Alice (2013). Evaluation of micronutrients in seeds of Pumpkin varieties grown by small land holder farmers in the Lake Victoria Basin. African Journal of Food Science and Technology, 10(10): 221-228.

22) Roza Martha Perez Gutierrez (2016). Review of Cucurbita pepo (Pumpkin) its Phytochemistry and Pharmacology. Medicinal chemistry, 2:1-4.

23) Dhiman K, Gupta A, Sharma DK, Gill NS, Goyal A (2012). A review on the medicinally important plants of the family Cucurbitaceae. Asian J Clin. Nutr, 4: 16-26.

24) Dubey A, Mishra N (2010). Antimicrobial Activity of some selected vegetables. IJABTP, 1(3): 995-999.

25) Sun J, Blaskovich MA, Jove R, Livingston SK, Coppola D (2005). Cucurbitacin Q: A selective STAT3 activation inhibitor with potent antitumor activity. Oncogene Journal 24:3236-3245.47

26) Saha P, Mazumder UK, Haldar PK, Bala A (2011). Evaluation of hepatoprotective activity of Cucurbita maxima aerial parts. Journal of Herbal Medicine and Toxicology, 5(1): 17-22p.

27) C. Madhavi Latha and Yalla Reddy Kolavali (2011). Evaluation of antihypertensive activity of cucurbita maxima. European journal of pharmaceutical and medical Research. Journal of Herbal Medicine and Toxicology, 5(1): 17-22.

28) Xanthopoulou MN, Nomikos T, Fragopoulou E, Antonopoulou S (2009). Antioxidant and lipoxygenase inhibitory activities of pumpkin seed extracts. Food Research International, 142: 641-646.

29) Craig WJ (1997). Phytochemicals: guardians of our health. Journal of the American Dietetic Association, 97(10):199–204.

30) Nair R, Kalariya T (2005). Anti-bacterial activity of some Indian medicinal flora. Journal of biochemistry, 29: 41-47p.